Prediction of Warning Level in Aircraft Accidents

Classification Techniques

Prediction of Warning Level in Aircraft Accidents
Classification Techniques

A Monograph

Arockia Christopher AB
Department of Master of Computer Applications
Rathinam Technical Campus
Coimbatore, Tamilnadu, India

CWP

This edition has been published by Central West Publishing PTY LTD, (ABN 13 683 898 722) Australia

For more information about the books published by Central West Publishing PTY LTD, please visit https://centralwestpublishing.com

Disclaimer

Every effort has been made by the publisher, editors and authors while preparing this book, however, no warranties are made regarding the accuracy and completeness of the content. The publisher, editors and authors disclaim without any limitation all warranties as well as any implied warranties about sales, along with fitness of the content for a particular purpose. Citation of any website and other information sources does not mean any endorsement from the publisher, editors and authors. For ascertaining the suitability of the contents contained herein for a particular lab or commercial use, consultation with the subject expert is needed. In addition, while using the information and methods contained herein, the practitioners and researchers need to be mindful for their own safety, along with the safety of others, including the professional parties and premises for whom they have professional responsibility. To the fullest extent of law, the publisher, editors and authors are not liable in all circumstances (special, incidental, and consequential) for any injury and/or damage to persons and property, along with any potential loss of profit and other commercial damages due to the use of any methods, products, guidelines, procedures contained in the material herein.

NATIONAL LIBRARY OF AUSTRALIA

A catalogue record for this book is available from the National Library of Australia

ISBN (print): 978-1-922617-65-1

Dedication

To my parents

Preface

This work focuses on evaluation of risk and safety in civil aviation industry. The aim of the work is to study the performance of different classification algorithms on accident reports of the Federal Aviation Administration (FAA) accident / incident data system database. The work analyses number of accident data records for all categories of aviation between the years of 1950 to 2012. The classification algorithms such as DT, KNN, SVM, NN and NB are used to predict the warning level of the component as the class attribute. The rules are proved in terms of their accuracy and these results are seen to be very meaningful. This study also proves that the NB classifiers will perform better than other classifiers.

Table of Contents

Abbreviations

NTSB	National Transportation Board
AFM	Aircraft Flight Manual
SMS	Safety Management Systems
DT	Decision Tree
k-NN	k-Nearest Neighbour
NB	Naive Bayes
SVM	Support Vector Machine
FAA	Federal Aviation Administration
VFR	Visual Flight Rules

CHAPTER 1

INTRODUCTION

1.1 An Overview of Data Mining Techniques

Data mining refers to extracting or mining knowledge from large amounts of data. In other words, Data mining is the science, art, and technology of discovering large and complex bodies of data in order to discover useful patterns. Theoreticians and practitioners are continually seeking improved techniques to make the process more efficient, cost-effective, and accurate. Many other terms carry a similar or slightly different meaning to data mining such as knowledge mining from data, knowledge extraction, data/pattern analysis data dredging.

Data mining is an iterative process within which progress is defined by discovery, through either automatic or manual methods. Data mining is most useful in an exploratory analysis scenario in which there are no predetermined notions about what willconstitute an "interesting" outcome. Data mining is the search for new, valuable, and nontrivial information in large volumes of data. It is a cooperative effort of humans and computers. Best results are achieved by balancing the knowledge of human experts in describing problems and goals with the search capabilities of computers.

Data mining treats as a synonym for another popularly used term, Knowledge Discovery from Data, or KDD. In others view data mining as simply an essential step in the process of knowledge discovery, in which intelligent methods are applied in order to extract data patterns.

Gregory Piatetsky-Shapiro coined the term "Knowledge Discovery in Databases" in 1989. However, the term 'data mining' became more popular in the business and press communities. Currently, Data Mining and Knowledge Discovery are used interchangeably.
Nowadays, data mining is used in almost all places where a large amount of data is stored and processed.

In practice, the two primary goals of data mining tend to be prediction and description. Prediction involves using some variables or fields in the data set to predict unknown or future values of other variables of interest. Description, on the other hand, focuses in finding patterns describing the data that can be interpreted by humans. Therefore, it is possible to put data mining activities into one of two categories.

- Predictive data mining, which produces the model of the systemdescribed by the given dataset, or
- Descriptive data mining, which produces new, nontrivial informationbased on the available dataset.

On the predictive end of the spectrum, the goal of data mining is to produce a model, expressed as an executable code, which can be used to perform classification, prediction, estimation, or other similar tasks.

On the other, descriptive, end of the spectrum, the goal is to gain an understanding of the analyzed system by uncovering pattern and relationships in large datasets. The relative importance of prediction and description for particular data mining application can vary considerably. The goals of prediction and description are achieved byusing data-mining techniques for the following data mining tasks:

- Classification – discovery of predictive learning function that classifies any data item into one of several predefined classes.
- Regression – discovery of predictive learning function, which maps any data item to a real-value prediction variable.
- Clustering – a common descriptive task in which one seeks to identify a finite set of categories or clusters to describe the data.
- Summarization – an additional descriptive task that involves methods forfinding a compact description for a set (or subset) of data.
- Dependency modelling – finding a local model that describes

significantdependencies between variables or between the values of a feature in a data set or in a part of a data set.

- Change and Deviation Detection – discovering the most significantchanges in the data set.

Data mining is a long process of research and product development. Thisevolution began when business data was first stored on computers, continued with improvements in data access, and more recently, generated technologies that allow users to navigate through their data in real time. Data mining takes this evolutionary process beyond retrospective data access and navigation to prospective and proactive informationdelivery. Data mining is readyfor application in the business sector because it is supportedby three technologies that are now sufficiently mature:

- Massive data collection
- Powerful multiprocessor computers
- Data mining algorithms

One of the greatest strengths of data mining is reflected in its wide range of methodologies and techniques that can be applied to a host of problem sets. Since data mining is a natural activity to be performed on large data sets, one of largest target marketsis the entire data warehousing, data-mart, and decision-support community, encompassingprofessionals from such industries as aircraft, retail, manufacturing, telecommunications, healthcare, insurance, and transportation.

Data Prediction is a two-step process, similar to that of data classification. Although, for prediction, we do not utilize the phrasing of “Class label attribute” because the attribute for which values are being predicted is consistently valued (ordered) instead of categorical (discrete-esteemed and unordered). The attribute can be referred to simply as the predicted attribute. Prediction can be viewed as the construction and use of a model to assess the class of an unlabeled object, or to assess the value or value ranges of an attribute that a given object is likely to have.

1.2 Data mining in various applications

Advantages of using data mining in various applications such as banking, manufacturing and production, marketing, health care etc., are as follows

- **Banking:** Data mining supports banking sector in the process of searching a large database to discover previously unknown patterns; automate the process of finding predictive information. Data mining helps to forecast levels of bad loans and fraudulent credit cards use, predicting credit card spending by new customers and predicting the kinds of customer best respond to new loan offered by the banks.
- **Manufacturing and production:** Data mining helps to predict the machine failures and finding key factors that control optimization of manufacturing capacity.
- **Marketing:** Data mining facilitates marketing sector by classifying customer demographic that can be used to predict which customer will respond to a mailing or buy a particular product and it is very much helpful in growth of business.
- **Health-Care:** Data mining supports a lot in health care sector. It supports health care sector by correlating demographics of patients with critical illnesses, developing better insights on symptoms and their causes and learning how to provide proper treatments.
- **Insurance:** Data mining assists insurance sector in predicting fraudulentclaims and medical coverage cost, classifying the important factors that affect medical coverage and predicting the customers' pattern which customer will buy new policies.
- **Law:** Law enforcement is helped by data mining by monitoring the behaviour patterns of the criminals. Tracking crime pattern, locations andcriminal behaviours, identifying various attributes to data mining, assist in solving criminal cases.
- **Government and Defence:** Data mining helps to forecast the cost of moving military equipment and predicting resource consumption. Apart from that it assists in testing strategies for potential military operations and improving

homeland security by mining data from many sources.

- **Brokerage and Securities trading:** Data mining assists in predicting the change in bond prices and forecasting the range of stock fluctuation determining the night time to buy or sell stocks.
- **Computer hardware and software:** Predicting disk-failures and potential security violations can be done by data mining.
- **Airlines:** It supports in checking the feasibility of adding routes to increase the business profit and to decrease the loss by capturing data on the location and the ultimate destination of passengers.

Data mining is highly useful in the following domains –

- Market Analysis and Management
- Corporate Analysis & Risk Management
- Fraud Detection
- Apart from these, data mining can also be used in the areas of production control, customer retention, science exploration, sports, astrology, and Internet Web Surf-Aid

Market Analysis and Management

Listed below are the various fields of market where data mining is used –

Customer Profiling – Data mining helps determine what kind of people buy what kind of products.

Identifying Customer Requirements – Data mining helps in identifying the best products for different customers. It uses prediction to find the factors that may attract new customers.

Cross Market Analysis – Data mining performs Association/correlations between product sales.

Target Marketing – Data mining helps to find clusters of model customers who share the same characteristics such as interests, spending habits, income, etc.

Determining Customer purchasing pattern – Data mining helps in determining customer purchasing pattern.

Providing Summary Information – Data mining provides us various multidimensional summary reports.

Corporate Analysis and Risk Management
Data mining is used in the following fields of the Corporate Sector.

Finance Planning and Asset Evaluation – It involves cash flow analysis and prediction, contingent claim analysis to evaluate assets.

Resource Planning – It involves summarizing and comparing the resources and spending.

Competition – It involves monitoring competitors and market directions.

Fraud Detection
Data mining is also used in the fields of credit card services and telecommunication to detect frauds. In fraud telephone calls, it helps to find the destination of the call, duration of the call, time of the day or week, etc. It also analyzes the patterns that deviate from expected norms.

1.3 Challenges in Data Mining

There are many challenges faced by the data mining and these challenges of datamining are pointed as follows

- Scalability
- Complex and Heterogeneous Data
- Network Setting
- Data Quality
- Data Ownership and Distribution
- Dimensionality
- Privacy preservation Streaming Data

According to National Transportation Board (NTSB) statistics, in the last 40 years, approximately 80 percent of aviation accidents have been caused by "pilot error." Many of these accidents are the result of the tendency to focus flight training on the physical aspects of flying the aircraft by teaching the student pilot enough aeronautical knowledge and skill to pass the written and practical tests.

Risk management is ignored, with sometimes fatal results. The certificated flight instructor (CFI) who integrates risk management into flight training teaches aspiring pilotshow to be more aware of potential risks in flying, how to clearly identify those risks, and how to manage them successfully.

"A key element of risk decision-making is determining if the risk is justified."

The risks involved with flying are quite different from those experienced in dailyactivities. Managing these risks requires a conscious effort and established standards (or amaximum risk threshold). Pilots who practice effective risk management havepredetermined personal standards and have formed habit patterns and checklists to incorporate them. The goal is to reduce the general aviation accident rate involving poor risk management. Pilots who make a habit of using risk management tools will find their flights considerably more enjoyable and less stressful for themselves and their passengers.In addition, some aircraft insurance companies reduce insurance rates after a pilot completes a formal risk management course.

Data mining techniques can be used to extract valuable insights and knowledge from large amounts of data. The advantages of data mining techniques include:

- **Increased efficiency:** Data mining techniques can help organizations to make better use of their data, by automating many of the processes that would otherwise have to be performed manually.
- **Improved decision making:** Data mining techniques can help organizations to make more informed decisions, by providing them with a more complete understanding of their data.

- **Enhanced competitiveness:** Data mining techniques can help organizations to gain a competitive edge by providing them with new insights and knowledge that they can use to improve their products, services, and operations.
- **Better customer service:** Data mining techniques can be used to identify patterns and trends in customer behavior, which can help organizations to improve their customer service and increase customer satisfaction.
- **Fraud detection**: Data mining techniques can be used to identify patterns and trends in financial data, which can help organizations to detect and prevent fraud.

However, data mining techniques also have some disadvantages, including:

- **Complexity:** Data mining techniques can be complex, requiring specialized knowledge and expertise to implement and interpret the results.
- **Privacy concerns:** Data mining techniques can raise privacy concerns, as they often involve the collection and analysis of sensitive personal information.
- **Ethical considerations:** Data mining can raise ethical considerations, as the results can be used to discriminate against certain groups or individuals.
- **Data quality**: The quality of the data can affect the results of data mining techniques, and poor quality data can lead to inaccurate or misleading results.
- **Computational cost:** Data mining techniques can be computationally expensive, requiring significant resources to process and analyze large amounts of data.
- **Overfitting**: Overfitting is a common problem in data mining, it occurs when a model is trained too well on the training data and performs poorly on new, unseen data.

This Risk Management makes available recommended tools for determining andassessing risk in order to make the safest possible flight with the least amount of risk.

Risk is the future impact of a hazard that is not controlled or eliminated. It can be viewed as future uncertainty created by the hazard. If it involves skill sets, the same situation may yield different risk.

- If a nick is not properly evaluated, the potential for propeller failure is unknown.
- If the aircraft is not properly bonded and grounded, there is a build-up ofstatic electricity that can and will seek the path of least resistance to ground. If the static discharge ignites the fuel vapor, an explosion may beimminent.
- A fatigued pilot is not able to perform at a level commensurate with the mission requirements.
- The owner of a homebuilt aircraft decides to use bolts from a local hardware store that cost less than the recommended hardware, but look the same and appear to be a perfect match, to attach and secure the aircraftwings. The potential for the wings to detach during flight is unknown.

In scenario 3, what level of risk does the fatigued pilot present? Is the risk equalin all scenarios and conditions? Probably not. For example, look at three different conditions in which the pilot could be flying:

- Day visual meteorological conditions (VMC) flying visual flight rules (VFR)
- Night VMC flying VFR
- Night instrument meteorological conditions (IMC) flying instrumentflight rules (IFR)

In these weather conditions, not only the mental acuity of the pilot but also the environment he or she operates within affects the risk level. For the relatively new pilot versus a highly experienced pilot, flying in weather, night experience, and familiarity withthe area are assessed differentlyto determine potential risk. For example, the experiencedpilot who typically flies at night may appear to be a low risk, but other factors such as fatigue could alter the risk assessment.

When coming to the condition of aircraft the bolts purchased at the hardware centre were simple low strength material bolts while the wing bolts specified by the manufacturer were close-tolerance bolts that were corrosion resistant. The bolts thehomebuilder used to attach the wings would probably fail under the stress of take-off.

The study of human behaviour is an attempt to explain how and why humans function the way they do. A complex topic, human behaviour, is a product both of innate human nature and of individual experience and environment. Definitions of human behaviour abound, depending on the field of study. In the scientific world, human behaviour is seen as the product of factors that cause people to act in predictable ways.

The Federal Aviation Administration (FAA) utilizes studies of human behaviourin an attempt to reduce human error in aviation. Historically, the term "pilot error" has beenused to describe an accident in which an action or decision made bythe pilot was the causeor a contributing factor that led to the accident. This definition also includes the pilot's failure to make the correct decision or take proper action. From a broader perspective, thephrase "human factors related" more aptly describes these accidents. A single decision or event does not lead to an accident, but a series of events; the resultant decisions together form a chain of events leading to an outcome. Many of these events involve the interactionof flight crews. In fact, airlines have long adopted programs for crew resource management(CRM) and line oriented flight training (LOFT) which has had a positive impact upon bothsafety and profit. These same processes can be applied (to an extent) to general aviation.

Human error may indicate where in the system a breakdown occurs, but it provides no guidance as to why it occurs. The effort of uncovering why pilots make mistakes is multidisciplinary in nature. In aviation and with pilots in particular some of thehuman factors to consider when examining the human role are decision-making, design of displays and controls, flight deck layout, communications, software, maps and charts, operating manuals, checklists and system procedures. Any one of the above could be or become a stress that triggers a breakdown in the human performance that results in a critical human error.

Since poor decision-making by pilots (human error) has been identified as amajor factor in many aviation accidents, human behaviour research tries to determine an individual's predisposition to taking risks and the level of an individual's involvement in accidents. Drawing upon decades of research, countless scientists have tried to figure out how to improve pilot performance.

A study in 1951 published by Elizabeth Mechem Fuller and Helen B. Baune of the University of Minnesota determined there were injury-prone children. The study comprised of two separate groups of second grade students. Fifty five students were considered accident repeaters and 48 students had no accidents. Both groups were from the same school of 600 and their family demographics were similar. The accident-free group showed a superior knowledge of safety and were considered industrious and cooperative with others but were not considered physically inclined.
The accident-repeater group had better gymnastic skills, were considered aggressive and impulsive, demonstrated rebellious behaviour when under stress, were poorlosers, and liked to be the center of attention. One interpretation of this data of an adult predisposition to injury stems from childhood behaviour and environment leads to the conclusion that any pilot group should be consist only of pilots who are safety conscious, industrious, and cooperative. Clearly, this is not only an inaccurate inference, but also an impossible target, since pilots are drawn from the general population and exhibit all typesof personality traits.

The Pilot's Health

One of the best ways pilots can mitigate risk is a self-evaluation to ensure they are in good health. A standardized method used in evaluating health employs the IMSAFEchecklist. It can easily and effectively be used to determine physical and mental readinessfor flying and to provide a good overall assessment of the pilot's well-being.

- Illness—Illness is an obvious pilot risk.

- Medication—any medicines that might affect pilot's judgment or make him/ her drowsy.
- Stress—under psychological pressure from the job. Money, health, or family problems

Stress causes concentration and performance problems.

While the regulations list medical conditions that require grounding, stress is notamong them. The pilot should consider the effects of stress on performance.

- Alcohol— As little as one ounce of liquor, one bottle of beer, or four ounces of wine can impair flying skills. Alcohol also renders a pilot moresusceptible to disorientation and hypoxia.
- Fatigue—Fatigue continues to be one of the most insidious hazards to flight safety, as it may not be apparent to a pilot until serious errors are made.
- Emotion—experienced any emotionally upsetting event.

Stress Management

Everyone is stressed to some degree almost all of the time. A certain amount ofstress is good since it keeps a person alert and prevents complacency. Effects of stress are cumulative and, if the pilot does not cope with them in an appropriate way, they can eventually add up to an intolerable burden. Performance generally increases with the onsetof stress, peaks, and then begins to fall off rapidly as stress levels exceed a person's abilityto cope. The ability to make effective decisions during flight can be impaired by stress.

There are two categories of stress acute and chronic. These are both explained in aero medical factors, of the pilot's handbook of aeronautical knowledge. Factors referredto as stress can affect decision-making skills and increase a pilot's risk of error in the flightdeck. For instance, imagine a cabin door that suddenly opens in flight on a Bonanza climbing through 1,500 feet on a clear sunny day. It may startle the pilot, but the stress would wane when it becomes, apparent that the situation is not a serious hazard. Yet, if thecabin door opened in instrument meteorological conditions (IMC), the stress level would be much higher despite little difference between the two scenarios. Therefore, one can conclude that our perception of problems (and the stress they create) is related to the environment in which the problems occur.

Another example is that mechanical problems always seem greater at night, a situation that all pilots have experienced. The key to stress management is to stop, think, and analyze before jumping to a conclusion. There is usually time to think before drawingconclusions. There are several techniques to help manage the accumulation of life stress, and prevent stress overload.
For example, to help reduce stress levels, set aside time for relaxation each day or maintain a program of physical fitness. To prevent stress overload,learn to manage time more effectively to avoid pressures imposed by getting behind schedule and not meeting deadlines.

Aircraft

Aircraft performance figures and the aircraft flight manual (AFM) are based ona new aircraft flown by a professional test pilot. There are other factors to keep in mind while assessing personal and aircraft performance. This aircraft uses the runways availablefor the trip with an adequate margin of safety under the conditions to be flown. For instance, consider an AFM for an aircraft that indicates a maximum demonstrated crosswindcomponent of 15 knots. This is the maximum crosswind that the manufacturer's test pilot demonstrated in the aircraft's certification.

Weather

Weather is a major environmental consideration. As pilots set their own personalminimums, they should evaluate the weather for a particular flight by considering the following:

In mountainous terrain, consider having higher minimums for ceiling andvisibility, particularly if the terrain is unfamiliar.

- Consider the possibility that the weather may be different from forecast.Have alternative plans and be ready and willing to divert should an unexpected change occur.
- Consider the winds at the airports being used and the strength of the crosswind component.
- If flying in mountainous terrain, consider whether there are strong windsaloft. Strong winds in mountainous terrain can cause severe turbulence and downdrafts and be very hazardous for aircraft even when there is noother significant weather.

Terrain

Evaluation of terrain is another important component of analyzing the flight environment.

- To avoid terrain and obstacles, especially at night or in low visibility, determine safe altitudes in advance by using the altitudes shown on visual flight rules (VFR) and instrument flight rules (IFR) charts during pre- flight planning.
- Use maximum elevation figures (MEF) and other easily obtainable data to minimize chances of an inflight collision with terrain or obstacles.

Airport

The lights are available at the destination and alternate airports (e.g., visual approach slope indicator (VASI), precision approach path indicator (PAPI) or instrument landing system (ILS), glideslope guidance

Check the Notices to Airmen (NOTAM) for closed runways or airports. Look for runway or beacon lights out, nearby towers, etc.

- Choose the flight route wisely. An engine failure gives the nearbyairportssupreme importance.
- There are shorter or obstructed fields at the destination

and/or alternate airports.

Airspace

If the trip is over remote areas, take appropriate clothing, water, and survival gear on-board in the event of a forced landing.
If the trip includes flying over water or unpopulated areas with the chance oflosing visual reference to the horizon, the pilot must be prepared to fly IFR.

Check the airspace and any temporary flight restrictions (TFRs) along the route of flight.

Night time

Night flying requires special consideration. If the trip includes flying at night over water or unpopulated areas with the chance of losing visual reference to the horizon,the pilot must be prepared to fly IFR.

The flight conditions allow a safe emergency landing at night. Pre-flight all aircraft lights, interior and exterior, for a night flight. Carry at least two flashlights one forexterior pre-flight and a smaller one that can be dimmed and kept nearby.

The human eye will see nothing outside that is dimmer than the flight deck lighting. Always fly at night with the interior lights as dim as possible. As the flight progresses and the eyes adjust to the darkness, usually the interior lights can be dimmed further, aiding the outside vision. If the interior lights will not dim, that would increase therisk factors by restricting the pilot's outside vision probably not the time for a night flight.

Visual Illusions

Although weather, terrain, airport conditions, and night versus daylight flying each produce unique challenges, together these factors conspire against a pilot's senses. Itis important to understand that unwittingly these factors can create visual illusions and cause spatial disorientation producing challenges the pilot did not anticipate.

Even the besttrained pilots sometimes fail to recognize a problem until it is too late to complete a flightsafely.

An accident involving a Piper PA-32 and an airline transport pilot illustrates howvisual illusions can create problems that lead to an accident. In this case, the aircraft collided with terrain during a landing. The sole occupant of the airplane was an airline transport pilot who was not injured.

The airplane owned and operated by the pilot, sustained substantial damage. The personal transportation flight was being operated in visual meteorological conditions (VMC) in mid-afternoon. Although it was not snowing, there was snow on the ground.

External Pressures

External pressures are influences which are external to the flight that create a sense of pressure to complete a flight often at the expense of safety. Factors that can be external pressures include the following:

Someone is waiting at the airport for theflight's arrival.A passenger the pilot does not want to disappoint.
The desire to demonstrate pilot skills

The desire to impress someone (Probably the two most dangerous words in aviation are "Watch this!")

Desire to satisfy a specific personal goal

A pilot's general goal-completion orientation

The emotional pressure associated with acknowledging that skill and experiencelevels may be lower than a pilot would like them to be.

The following accident offers an example of how external pressures influence a pilot. Two pilots were giving helicopter demonstrations at an air show. The first pilot demonstrated a barrel roll in front of the stands. Not to be outdone, the second pilot with passengers decided to execute a hammerhead type maneuver. Flying past the stands at 90 knots, the pilot pulled the helicopter into a steep climb that ended at about 200 feet. Whenthe speed dissipated to near zero, he rolled back to the ground in a nose-low attitude to regain airspeed with the obvious intention of pulling the aircraft out of the dive near the ground.

An error in judgment led to the pilot being unable to pull the helicopter out of the dive. The helicopter struck the ground, killing all onboard. The desire to impress someone can be a powerful external pressure, especially when coupled with the internal pressure of pride. Perhaps the pilot decided to perform a maneuver not in his training profile, or one in which he had not demonstrated proficiency. It appears there was nothingin this pilot's experiences to help him effectively access the high risk of this maneuver in an aircraft loaded with passengers. It is not uncommon to see people motivated by externalpressures and driven internally by their own attitude.

Management of external pressure is the single most important key to riskmanagement because it is the one risk factor category that can cause a pilot to ignore all other risk factors. External pressures place time-related pressure on the pilot and figure intoa majority of accidents. Helicopter Emergency Medical Service (HEMS) operations, unique due to the emergency nature of the mission, are an example of how external pressures influence pilots. Emergency medical services (EMS) pilots often ferry criticallyill patients, and the pilot is driven by goal completion.

In order to reduce the effect of this pressure, many EMS operators do not notifyto the EMS pilot of the prospective patient's condition, but merely confine the location ofthe patient pick/up and restrict the pilot's decision-making role to the response to the question "Can the pick/up and transportation to the medical care center be made safely?" Risking three or four lives in an attempt to save one life is not a safe practice. The use of personal standard operating procedures (SOPs) is one way to manage external pressures. The goal is to supply a release for the external pressures of a flight. These procedures include, but are not limited to:

Allow time on a trip for an extra fuel stop or to make an unexpected landing because of weather.

Have alternate plans for a late arrival or make backup airline reservations for must-be-there trips.

For really important trips, plan to leave early enough so that there would still betime to fly to the destination.

Advise those who are waiting at the destination that the arrival may be delayed.
Know how to notify them when delays are encountered.

Manage passenger expectations. Ensure passengers know that they might not arrive on a firm schedule, and if they must arrive by a certain time, they should make alternative plans.

Eliminate pressure to return home, even on a casual day flight, by carrying a small overnight kit containing prescriptions, contact lens solutions, toiletries, or other necessities on every flight.

The key to managing external pressure is to be ready for and accept delays. Remember that people get delayed when travelling on airlines, driving a car, or taking a bus. The pilot's goal is to manage risk, not increase it.

Safety plays a paramount role in aviation industry. Safety in aviation industry turns into a major issue in numerous nations today. Truly, aviation safety has been based on the reactive investigation of past mishaps and the acquaintance of restorative movements with avert the repeat of the aforementioned occasions. Historically, aviation safety has been built upon the reactive analysis of past accidents and the introduction of corrective actionsto prevent the recurrence of those events.

With today's extremely low accident rate, it is increasingly difficult to make further improvements to the level of safety by using this approach. Therefore, a proactive approach to managing safety that concentrates on the control of processes has been developed solely relying on inspection and remedial actions on end products. This innovation in aviation system safety is called a Safety Management System (SMS), an expression indicating that safety efforts are most effective when made a fully integrated part of the business operation. It is now generally accepted that most aviation accidents result from human error.

It would be easy to conclude that these errors indicate carelessness or incompetence on the job, but that would not be accurate. Investigations have found that thehuman is the only the last link in a chain that leads to an accident. These accidents will notbe prevented by merely changing people.

Increased safety can only occur when the underlying causal factors are addressed. The aircraft accidents have many reasons. Many factors influencing aircraft have been found out by many researchers after deep study of many major aircraft accidents. These factors may be of internal factors of the aircraft or external conditions of the environment in which the aircraft operates. Though we cannot completely avoid the accidents we can reduce it by carefully identifying the factors resulting in it and take precautions to avoid the accidents.

To maintain the current low level of air accident fatalities aviation safety is a must. It is important to advance from a reactive framework where regulations are changedas a consequence of experience towards ace dynamic framework which endeavours to suspect potential safety hazards so as to decrease the probability of a mishap.
Much of airplane design involves ensuring a high level of safety for every airplane; however, crashes do occur, and the resulting loss of life is tragic. Determining thecause of a crash can improve the overall aviation safety by bringing to light any faults in the current design of an aircraft, and it can serve as a guide for improving future designs. The key problem lies in deciding which of many factors were the primary causes of any single crash. Often the expert crash investigators disagree.

Generally, the more data that is recorded and available on a flight, the more investigators are able to piece together what happened. For instance, radar tracking data, weather data, a Cockpit Voice Recorder (CVR), and a Flight Data Recorder (FDR) are allimportant tools for an aircraft crash investigator. Unfortunately, only a limited amount of data is recoverable from most accident sites because of the extent of the damage done to the plane.

It is the role of the investigator to determine the most likely scenario with whatinformation he has. For many commuter airplanes, the CVR is the investigators only insight into the mind of the flight crew. In fact, even if the crew does not specifically saywhat the problem is, the cause of the crash can sometimes be determined from a thoroughanalysis of the CVR.

In recent years, researchers like Ronald Stearman at the University of Texas at Austin have begun to use the non-voice signatures present on the recordings to provide insight into the cause of a crash. Physical events such as the failure of a load-bearing member propagate both through the air in acoustic waves, as well as through materials in the form of pressure changes. Stearman has demonstrated that CVRs are sensitive to boththe acoustic and non-acoustic spectrums. The purpose of this report is to analyze the signals present on the CVR from one particular crash and to determine if any signals are the result of events that contributed to the accident, but were unknown to the crew.

Unfortunately, however, airplane pilots and their passengers are not immune toerror, and, as a result, disastrous situations involving injury or death to innocent people dooccur. Because of the tremendous forces involved in airplane crashes, survivors often suffer catastrophic injury, including traumatic brain injury, spinal cord injury, amputation,and severe burn injuries.

Any airplane accident, whether it involves a private or commercial airplane, requires immediate and thorough investigation. Typically, this will involve investigators from the FAA. Even though the government usually investigates most airplane crashes thoroughly, the families of those sustaining injury or death should promptly seek competent legal advice from attorneys experienced in this area of the law.

The air transport industry plays a major role in world economic activity. One ofthe key elements for maintaining the vitality of civil aviation is to ensure safe, secure, efficient and environmentally sustainable operations at the global, regional and national levels.

A specialized agency of the United Nations, the International Civil Aviation Organization (ICAO) was created in 1944 to promote the safe and orderly development ofinternational civil aviation throughout the world. ICAO sets the Standards and Recommended Practices (SARPs) necessary for aviation safety, security, efficiency and environmental protection on a global basis. ICAO serves as the primary forum forcooperation in all fields of civil aviation among its 191 Member States. Improving the safety of the global air transport system is ICAO's guiding and most fundamental strategicobjective. The Organization works constantly to address and enhance global aviation safety through the following coordinated activities:

In every case, these activities are augmented by ICAO's detailed appraisal of global and regional aviation safety metrics on the basis of established risk management principles—a core component of contemporary State Safety Programmes (SSP) and Safety Management Systems (SMS). Applying these principles in the field of aviation safety requires ICAO to pursue a strategy comprised of proactive and reactive safety analysis andrisk management processes.

In all of its coordinated safety activities, ICAO strives to achieve a balance between assessed risk and the requirements of practical, achievable and effective risk mitigation strategies. The ICAO Safety Report is published annually each year in electronic format to provide updates on safety indicators including accidents and related risk factorsoccurring in the previous year.

In addition, ICAO publishes a triennial Special Edition – the State of Global Aviation Safety report – during August of each year in which an ICAO Assembly is held.The State of Global Aviation Safety reports provide Member States, the aviation community and the travelling public with a high-level analysis of air transport safety trends and indicators. They also include comprehensive accounts of the significant aviation safetyprogrammes being undertaken by ICAO and its partners, highlighting the organization's important leadership role in fostering increased cooperation and innovation to enhance airtransport safety outcomes worldwide. The State of Global Aviation Safety Report is published in six languages, electronically and in print.

The aviation accident database includes:

All civil and commercial aviation accidents of scheduled and non-scheduled passenger airliners worldwide, which resulted in a fatality (including all U.S. Part 121 andPart 135 fatal accidents)

All cargo, positioning, ferry and test flight fatal accidentsAll military transport accidents with 10 or more fatalities
All commercial and military helicopter accidents with more than 10 fatalities

All civil and military airship accidents involving fatalitiesAviation accidents involving the death of famous people Aviation accidents or incidents of noteworthy interest.

Data mining involves the use of data analysis tools to discover previously unknown, valid patterns and relationships from large amounts of data stored in databases,data warehouses, or other information repositories. Feature selection is an essential pre- processing step of data mining that helps increase the predictive performance of a model. The main aim of feature selection is to choose a subset of features with high predictive information and eliminate irrelevant features with little or no predictive information.

Classification is a data mining technique used to classify or predict number of individual collectively associated for data instances. The object of classification is to exactly predict the objective class for each case in the dataset. One of the commendable features of classifier is its ability to tolerate noise. Its difficulty lies in handling quantitativedata appropriately. Risk and safety have always been essential applications in Aviation Company. Aviation accidents may result in human injury or even death. An airline company collects several case reports including structural and textual data. Institutions make important investment in order to collect such information from several sources.

In machine learning and statistics, feature selection, also known as variable selection, attribute selection or variable subset selection, is the process of selecting a subsetof relevant features for use in model construction. The central assumption when using a feature selection technique is that the data contains many redundant or irrelevant features.Redundant features are those which provide no more information than the currently selected features, and irrelevant features provide no useful information in any context. Feature selection techniques provide three main benefits when constructing predictive models: improved model interpretability, shorter training times and enhanced generalization by reducing over fitting. Feature selection is also useful as part of the data analysis process, that shows which features are important for prediction, and how these features are related.

We have used different feature selection techniques for discovering attributes evaluator and generating a classification algorithm. These techniques are Correlation Feature Selection Subset Evaluator, Consistency Subset Evaluator, Gain Ratio feature evaluator, Information Gain Attribute Evaluator, OneR feature evaluator, Principal Components Attribute Transformer, ReliefF Attribute Evaluator and Symmetrical Uncertainty Attribute.

In this work, we have applied different data mining methods on the incident reports. We find feature selection between attributes and classification algorithms to find the relations and rules about the incidents resulted in fatality. We also applied Pre- processing methods to select the attributes which will be used in the analysis. This work focuses on different feature selection techniques and applying traditional classification algorithms to understand and clean the dataset.

Feature selection methods are applied in order to reduce the number of initial features. The classification data mining is used to predict the warning level of thecomponent as the class attribute. The number of input attributes can be reduced by the feature selection methods and output of the reduction mechanism is combined with variable input classification algorithm. The proposed model will be able to predict warning level ofthe accident for airline companies. We have explored the use of the different classificationtechniques on aviation components data. Based on the analysis of the aircraft accident dataof this report we recommend some of the safety and improvement of the competitiveness of the aviation company process within this sector.

In this work, five classification algorithms Decision Tree (DT), Naive Bayes Classification (NBC), Artificial Neural Network (ANN), K-Nearest Neighbour (KNN) and Support Vector Machines (SVM) have been considered for comparing their performance based on the aircraft data. In this experimentation, 10-fold cross-validation method is usedas the test option. This work concentrates on performance of classification algorithms withdifferent features combinations such as CFS, Consistency, GR, Information Gain, OneR, PCA, ReliefF and Symmetric with the selected datasets and also to improve the classification accuracy.

Data Mining is an analytic process which designed to explore data usually largeamounts of data. Data Mining is often considered to be "a blend of statistics. We have usedfive data mining techniques for discovering classification rules and generating a decision tree. These techniques are DT, KNN, NB, NN and SVM. Data mining tools WEKA is usedin this work. This is particularly so under current conditions of continuous growth in air transport demand, frequent scarcity of airport and infrastructure capacity, and thus permanent and increased pressure on the system components. There is also the growing public and operators' awareness of these and other system externalities such as airpollution, noise, land use, water/soil pollution and waste management, and congestion. Risk and safety have always been essential considerations in aviation.

With the rapid growth in air travel, flight delays, cancellations and incidents/accidents have also dramatically increased in recent years. Aviation accidents may result in human injury or even death. An airline company collects several case reportsincluding structural and textual data. Institutions make important investment in order to collect such information from several sources. Databases' storing this information keeps essential data; besides there exit some crucial problems in analyzing the information.

In this work, we applied different data mining methods on the incident reports. We, the decision trees and classification algorithms to find the relations and rules about theincidents resulted in fatality. The decision tree techniques of data mining is use to predict the warning level of the component as the class attribute. All data sets reduce by find dependencies are more effective and give the rules that are finding by these datasets. New and Novel classification model will be able to predict warning level of the accident for airline companies. We have explored the use of the decision tree techniques on aviation components data. We gave this rules that were found by this analysis for the experts of Aviation Company. Some safety recommendations are address to the Airline Aviation Administration.

The work is organized as follows. In Sections 2 and 3 we mentioned the background and data mining methods of data mining. In Section 4 we described the datasetdescriptions about classification rule discovery for the aviation incidents resulted fatality.

CHAPTER 2

LITERATURE SURVEY

2.1 Background work

Data mining tools are then used to uncover useful patterns and relationships from the data captured. Data mining methods have been successfully introduced in many fields.It is still a research topic, but besides that of tremendous interest or the industry in order tosolve their real-world problems. Within this frame, the decision tree classification providesa rapid and effective method of categorizing datasets. For many problems of classificationwhere large datasets are used and the information contained is complex and may contain errors, decision trees provide a useful solution. Aitken head presented a method which combines the decision tree paradigm with different evolutionary concepts to produce a classification methodology.

Now we will mention on some data mining applications about different fields. Chang and Leu [1] proposed a data mining model and procedure to relate influence variables to project profitability. Chang and Chen [3] examine data mining technology to fulfill readers' requirements. Kirkos et al. [6] explored the effectiveness of data mining classification techniques in detecting firms that issue fraudulent financial statements and deals with the identification of factors associated to fraudulent financial statements. Mai etal. [4] applied data mining to forest data to determine the factors responsible for deforestation. Tso and Yau [13] presented three modeling techniques-regression analysis,decision tree and neural network-for the prediction of electricity energy consumption. Hsiaet al. [20] used data mining techniques to analyze the course references and course completion rates of enrollees in extension education courses at a university in Taiwan. Also in the aviation industry, data mining applications has been done. Bineid and Fielding [16]used data mining techniques to explain the development of a dispatch reliability prediction method for passenger aircraft. Nazeri and Zhang [18] described the application of data mining to analyzing severe weather impacts on National Airspace System performance.

Epidemiologists and mine safety researchers have still used it to perform many analyses, helping to guide research and best practice. Shyur [12] proposed a model that allows investigation of nonlinear effects of aviation safety factors and flexible assessmentof aviation risk.

Decision tree learning is a method commonly used in data mining. The goal is to create a model that predicts the value of a target variable based on several input variables. An example is shown on the right. Each interior node corresponds to one of the input variables; there are edges to children for each of the possible values of that input variable.Each leaf represents a value of the target variable given the values of the input variables represented by the path from the root to the leaf.

In pattern recognition, the k-nearest neighbour algorithm (k-NN) is a method forclassifying objects based on closest training examples in the feature space. K-NN is a typeof instance-based learning, or lazy learning where the function is only approximated locally and all computation is deferred until classification. The k-nearest neighbour algorithm is amongst the simplest of all machine learning algorithms: an object is classified by a majority vote of its neighbours, with the object being assigned to the class most common amongst its k nearest neighbours (k is a positive integer, typically small). If k = 1, then theobject is simply assigned to the class of its nearest neighbour.

Association analysis is the finding of association rules showing attribute-value conditions that occur frequently together in a given set of data. Association analysis is widely used for a market basket or transaction data analysis. Association rule mining is a significant and exceptionally dynamic area of data mining research. One method of association-based classification, called associative classification, consists of two steps. In the main step, association instructions are generated using a modified version of the standard association rule mining algorithm known as Apriori. The second step constructs a classifier based on the association rules discovered.

The Bayesian Classification represents a supervised learning method as well asa statistical method for classification. Assumes an underlying probabilistic model and it allows us to capture uncertainty about the model in a principled way by determining probabilities of the outcomes. It can solve diagnostic and predictive problems. ThisClassification is named after Thomas Bayes (1702-1761), who proposed the Bayes Theorem. Bayesian classification provides practical learning algorithms and prior knowledge and observed data can be combined. Bayesian Classification provides a usefulperspective for understanding and evaluating many learning algorithms. It calculatesexplicit probabilities for hypothesis and it is robust to noise in input data.

Classification is the processing of finding a set of models (or functions) that describe and distinguish data classes or concepts, for the purpose of being able to use the model to predict the class of objects whose class label is unknown. The determined model depends on the investigation of a set of training data information (i.e. data objects whose class label is known). The derived model may be represented in various forms, such as classification (if – then) rules, decision trees, and neural networks. Data Mining has a different type of classifier:

- Decision Tree
- SVM (Support Vector Machine)
- Generalized Linear Models
- Bayesian classification:
- Classification by Backpropagation
- K-NN Classifier
- Rule-Based Classification
- Frequent-Pattern Based Classification
- Rough set theory
- Fuzzy Logic

Decision Trees: A decision tree is a flow-chart-like tree structure, where each node represents a test on an attribute value, each branch denotes an outcome of a test, and tree leaves represent classes or class distributions. Decision trees can be easily transformed into classification rules. Decision tree enlistment is a nonparametric methodology for building classification models. In other words, it does not require any prior assumptions regarding the type of probability distribution satisfied by the class and other attributes. Decision trees, especially smaller size trees, are relatively easy to interpret. The accuracies of the trees are also comparable to two other classification techniques for a much simple data set. These provide an expressive representation for learning discrete-valued functions. However, they do not simplify well to certain types of Boolean problems.

Support Vector Machine (SVM) Classifier Method: Support Vector Machines is a supervised learning strategy used for classification and additionally used for regression. When the output of the support vector machine is a continuous value, the learning methodology is claimed to perform regression; and once the learning methodology will predict a category label of the input object, it's known as classification. The independent variables could or could not be quantitative. Kernel equations are functions that transform linearly non-separable information in one domain into another domain wherever the instances become linearly divisible. Kernel equations are also linear, quadratic, Gaussian, or anything that achieves this specific purpose.

A linear classification technique may be a classifier that uses a linear function of its inputs to base its decision on. Applying the kernel equations arranges the information instances in such a way at intervals in the multi-dimensional space, that there is a hyper-plane that separates knowledge instances of one kind from those of another. The advantage of Support Vector Machines is that they will make use of certain kernels to transform the problem, such we are able to apply linear classification techniques to nonlinear knowledge. Once we manage to divide the information into two different classes our aim is to include the most effective hyper-plane to separate two kinds of instances.

Generalized Linear Models: Generalized Linear Models (GLM) is a statistical technique, for linear modeling.GLM provides extensive coefficient statistics and model statistics, as well as row diagnostics. It also supports confidence bounds.

Bayesian Classification: Bayesian classifier is a statistical classifier.

They can predict class membership probabilities, for instance, the probability that a given sample belongs to a particular class. Bayesian classification is created on the Bayes theorem. Studies comparing the classification algorithms have found a simple Bayesian classifier known as the naive Bayesian classifier to be comparable in performance with decision tree and neural network classifiers. Bayesian classifiers have also displayed high accuracy and speed when applied to large databases. Naive Bayesian classifiers adopt that the exact attribute value on a given class is independent of the values of the other attributes. This assumption is termed class conditional independence. It is made to simplify the calculations involved, and is considered “naive”. Bayesian belief networks are graphical replicas, which unlike naive Bayesian classifiers allow the depiction of dependencies among subsets of attributes. Bayesian belief can also be utilized for classification.

Classification By Backpropagation: A Backpropagation learns by iteratively processing a set of training samples, comparing the network’s estimate for each sample with the actual known class label. For each training sample, weights are modified to minimize the mean squared error between the network’s prediction and the actual class. These changes are made in the “backward” direction, i.e., from the output layer, through each concealed layer down to the first hidden layer (hence the name backpropagation). Although it is not guaranteed, in general, the weights will finally converge, and the knowledge process stops.

K-Nearest Neighbor (K-NN) Classifier Method: The k-nearest neighbor (K-NN) classifier is taken into account as an example-based classifier, which means that the training documents are used for comparison instead of an exact class illustration, like the class profiles utilized by other classifiers. As such, there’s no real training section. once a new document has to be classified, the k most similar documents (neighbors) are found and if a large enough proportion of them are allotted to a precise class, the new document is also appointed to the present class, otherwise not. Additionally, finding the closest neighbors is quickened using traditional classification strategies.

Rule-Based Classification: Rule-Based classification represent the knowledge in the form of If-Then rules. An assessment of a rule evaluated according to the accuracy and coverage of the classifier. If more than one rule is triggered then we need to conflict resolution in rule-based classification. Conflict resolution can be performed on three different parameters: Size ordering, Class-Based ordering, and rule-based ordering. There are some advantages of Rule-based classifier like:

Rules are easier to understand than a large tree.

Rules are mutually exclusive and exhaustive.

Each attribute-value pair along a path forms conjunction: each leaf holds the class prediction.

Frequent-Pattern Based Classification: Frequent pattern discovery (or FP discovery, FP mining, or Frequent item set mining) is part of data mining. It describes the task of finding the most frequent and relevant patterns in large datasets. The idea was first presented for mining transaction databases. Frequent patterns are defined as subsets (item sets, subsequences, or substructures) that appear in a data set with a frequency no less than a user-specified or auto-determined threshold.

Rough Set Theory: Rough set theory can be used for classification to discover structural relationships within imprecise or noisy data. It applies to discrete-valued features. Continuous-valued attributes must therefore be discrete prior to their use. Rough set theory is based on the establishment of equivalence classes within the given training data. All the data samples forming a similarity class are indiscernible, that is, the samples are equal with respect to the attributes describing the data. Rough sets can also be used for feature reduction (where attributes that do not contribute towards the classification of the given training data can be identified and removed), and relevance analysis (where the contribution or significance of each attribute is assessed with respect to the classification task). The problem of finding the minimal subsets (redacts) of attributes that can describe all the concepts in the given data set is NP-hard.

However, algorithms to decrease the computation intensity have been proposed. In one method, for example, a discernibility matrix is used which stores the differences between attribute values for each pair of data samples. Rather than pointed on the entire training set, the matrix is instead searched to detect redundant attributes.

Fuzzy-Logic: Rule-based systems for classification have the disadvantage that they involve sharp cut-offs for continuous attributes. Fuzzy Logic is valuable for data mining frameworks performing grouping /classification. It provides the benefit of working at a high level of abstraction. In general, the usage of fuzzy logic in rule-based systems involves the following:

Attribute values are changed to fuzzy values.

For a given new data set /example, more than one fuzzy rule may apply. Every applicable rule contributes a vote for membership in the categories. Typically, the truth values for each projected category are summed.

Unlike classification and prediction, which analyze class-labeled data objects or attributes, clustering analyzes data objects without consulting an identified class label. In general, the class labels do not exist in the training data simply because they are not known to begin with. Clustering can be used to generate these labels. The objects are clustered based on the principle of maximizing the intraclass similarity and minimizing the interclass similarity. That is, clusters of objects are created so that objects inside a cluster have high similarity in contrast with each other, but are different objects in other clusters. Each Cluster that is generated can be seen as a class of objects, from which rules can be inferred. Clustering can also facilitate classification formation, that is, the organization of observations into a hierarchy of classes that group similar events together. A multilayer perceptron is a feed forward artificial neural network model that maps sets of input data onto a set of appropriate output. An MLP consists of multiple layersof nodes in a directed graph, with each layer fully connected to the next one. Except for the input nodes, each node is a neuron (or processing element) with a nonlinear activationfunction. MLP utilizes a supervised learning technique called back propagation for trainingthe network. MLP is a modification of the standard linear perceptron and can distinguish data that is not linearly separable.

Regression can be defined as a statistical modeling method in which previously obtained data is used to predicting a continuous quantity for new observations. This classifier is also known as the Continuous Value Classifier. There are two types of regression models: Linear regression and multiple linear regression models.

In machine learning, support vector machines (SVMs, also support vector networks) are supervised learning models with associated learning algorithms that analyzedata and recognize patterns, used for classification and regression analysis. The basic SVMtakes a set of input data and predicts, for each given input, which of two possible classes forms the output, making it a non-probabilistic binary linear classifier. Given a set of training examples, each marked as belonging to one of two categories, an SVM training algorithm builds a model that assigns new examples into one category or the other. An SVM model is a representation of the examples as points in space, mapped so that the examples of the separate categories are divided by a clear gap that is as wide as possible.

An artificial neural network (ANN) also referred to as simply a "Neural Network" (NN), could be a process model supported by biological neural networks. It consists of an interconnected collection of artificial neurons. A neural network is a set of connected input/output units where each connection has a weight associated with it. During the knowledge phase, the network acquires by adjusting the weights to be able to predict the correct class label of the input samples. Neural network learning is also denoted as connectionist learning due to the connections between units. Neural networks involve long training times and are therefore more appropriate for applications where this is feasible. They require a number of parameters that are typically best determined empirically, such as the network topology or "structure". Neural networks have been criticized for their poor interpretability since it is difficult for humans to take the symbolic meaning behind the learned weights. These features firstly made neural networks less desirable for data mining.

The advantages of neural networks, however, contain their high tolerance to noisy data as well as their ability to classify patterns on which they have not been trained. In addition, several algorithms have newly been developed for the extraction of rules from trained neural networks. These issues contribute to the usefulness of neural networks for classification in data mining.

An artificial neural network is an adjective system that changes its structure-supported information that flows through the artificial network during a learning section. The ANN relies on the principle of learning by example. There are two classical types of neural networks, perceptron and also multilayer perceptron.

CHAPTER 3

DATA MINING TECHNIQUES

3.1 Data Mining Methods

Data mining is the process of selection, exploration and modeling of large quantities of data to discover regularities or relations that are at first unknown with the aimof obtaining clear and useful results for the owner of the database. Data mining functionalities are used to specify the kind of patterns to be found in data mining tasks.

Classification of data is very important task in data mining and machine learning. The goal of classification is to correctly predict the value of a designated discrete class variable, given a vector of predictors or attributes.

The input to the problem is a dataset called the training set, which consists of a number of examples each having a number of attributes. The attributes are either continuous, when the attribute values are ordered, or categorical when the attribute valuesare unordered. One of the categorical attributes is called the class label or the classifying attribute.

The objective is to use the training set to build a model of the class label based on the other attributes such that the model can be used to classify new data not from the training dataset. Classification has been studied extensively in statistics, machine learning,neural networks and expert systems over decades. There are several classification methods:

Decision tree algorithmsBayesian algorithms Rule based algorithms Neural networks Support vector machines
Associative classification Distance based methods Genetic Algorithms
After selecting the most relevant features, we have applied different classification algorithms such as DT, NB, SVM, KNN and NN and per-

formance of these algorithms are then used to compare the effectiveness of various feature selection methods.

Data Mining is an analytic process which is designed to explore data-usually large amounts of data. We have used several classification techniques for discovering classification rules and generating a model. These techniques are DT, KNN, NB, NN and SVM. Aviation accidents may result in human injury or even death. An airline company collects several case reports including structural and textual data. In this study, we have applied different classification methods on the incident reports.

The decision tree model is used to predict the warning level of the component asthe class attribute. The proposed classification model will be able to predict warning levelof the accident for airline companies. We have explored the use of the decision tree techniques on aviation components data. Some safety recommendations are addressed to the Airline Aviation Administration based on the rules that were generated through this analysis.

3.2 Classification and prediction

Databases are rich with hidden information that can be used for making intelligent business decisions. Classification and prediction are two forms of data analysiswhich can be used to extract models describing important data classes or to predict futuredata trends. Whereas classification predicts categorical labels (or discrete values), prediction models continuous-valued functions. They may come from an experiment in which a sample of the entire database is tested in the real world and the results used to create a classifier. Classification is the process of finding a model that describes the data classes or concepts. The purpose is to be able to use this model to predict the class of objects whose class label is unknown. This derived model is based on the analysis of sets of training data. The derived model can be presented in the following forms:

Classification (IF-THEN) Rules
Decision Trees
Mathematical Formulae

Neural Networks

The list of functions involved in these processes are as follows:

Classification – It predicts the class of objects whose class label is unknown. Its objective is to find a derived model that describes and distinguishes data classes or concepts. The Derived Model is based on the analysis set of training data i.e. the data object whose class label is well known.

Prediction – It is used to predict missing or unavailable numerical data values rather than class labels. Regression Analysis is generally used for prediction. Prediction can also be used for identification of distribution trends based on available data.

Outlier Analysis – Outliers may be defined as the data objects that do not comply with the general behavior or model of the data available.

Evolution Analysis – Evolution analysis refers to the description and model regularities or trends for objects whose behavior changes over time.

Data classification is a two-step process. In the first step, a model is built describing a predetermined set of data classes or concepts. The model is constructed by analyzing database tuples described by attributes. Each tuple is assumed to belong to a predefined class, as determined by one of the attributes, called the class label attribute. In the context of classification, data tuples are also referred to as samples, examples, or objects. The data tuples analyzed to build the model collectivelyform the training data set.The rules can be used to categorize future data samples, as well as provide a better understanding of the database contents. In the second step, the model is used for classification. First, the predictive accuracy of the model (or classifier) is estimated. The holdout method is a simple technique which uses a test set of class-labeled samples. Thesesamples are randomly selected and are independent of the training samples. The accuracyof a model on a given test set is the percentage of test set samples that are correctly classified by the model.

If the accuracy of the model is considered acceptable, the model can be used toclassify future data tuples or objects for which the class label is not known. How is prediction different from classification? Prediction can be viewed as the construction and use of a model to assess the class of an unlabeled object, or to assess the value or value ranges of an attribute that a given object is likely to have. In our view, however, we refer to the use of predication to predict class labels as classification and the use of predicationto predict continuous values as prediction. This view is commonly accepted in data mining.

3.3 Classification by DT

Decision tree (DT) induction is a very popular and practical approach for patternclassification. Decision tree is constructed generally in a greedy, top down recursive manner. The tree can be constructed in a breadth first manner or depth first manner.

Decision tree structure consists of a root node, internal nodes and leaf nodes. The classification rules are derived from the decision tree and these rules are used to classify the records with unknown value for class label. The decision tree is constructed intwo phases: Building Phase and Pruning Phase.

In Building Phase of the tree the relevant attributes are selected based on attribute selection measures such as Information gain, Gain Ratio, Gini Index, etc. Once the relevant attribute is selected then the tree is constructed with that node as the root nodeand the distinct values of the attribute are denoted as branches. The process of selecting relevant attribute and representing the distinct values as branches are repeated until all theinstances in the training set belong to the same class label.

In Pruning Phase the sub-trees are eliminated which may over fit the data. Thisenhances the accuracy of a classification tree. Decision trees handle continuous and discrete attributes. Decision trees are widely used because they provide human readable rules, easy to understand, construction of decision tree is fast and it yields better accuracy.There are several algorithms to classify the data using decision trees.

C4.5 is the improved versions of ID3 algorithms or the output of C4.5 is the Decision tree. A Decision tree is similar to the tree structure having root node, intermediatenodes and leaf node. Each node in the tree consist of a decision and that decision leads to our result. Decision tree divides the input space of a dataset into mutually exclusive areas,each area having a label, a value or an action to describe its data points. Splitting criterionis used to calculate which attribute is the best to split that portion tree of the training data that reaches a particular node.

Decision tree can produce a model with rules that are human-readable and interpretable. The classification task using decision tree technique can be performed without complicated computations and the technique can be used for both continuous andcategorical variables. This technique is suitable for predicting categorical outcomes.

Data mining approaches have been successfully applied in different fields. Riskand safety have always been important considerations in aviation. There is a large amountof knowledge and data accumulation in aviation industry. These data can be stored in the form of pilot reports, maintenance reports, accident reports or delay reports. This work hasapplied the decision tree model on accident reports of the Federal Aviation Administration(FAA) accident or incident Data System database, which contains number of accident datainstances for all categories of aviation between the years of 1970 to 2012. The decision treeclassifier is used to predict the warning level of the component as the class attribute. We have explored the use of the decision tree technique on aviation components data. DecisionTree induction algorithm is applied to generate the model and the generated model is usedto predict the warning of accidents in the airline database. This work may be useful for Aviation Company to make better prediction.

A decision tree is a flow-chart-like tree structure, where each internal node denotes a test on an attribute, each branch represents an outcome of the test, and leaf nodesrepresent classes or class distributions. The topmost node in a tree is the root node.

In order to classify an unknown sample, the attribute values of the sample are tested against the decision tree. A path is traced from the root to a leaf node which holds the class prediction for that sample. Decision trees can easily be converted to classificationrules.

3.4 Classification by KNN

A k-Nearest Neighbor (k-NN) is one of the supervised learning algorithms that have been used in many applications in the field of data mining, statistical pattern recognition and many others. It follows a method for classifying objects based on the closest training examples in the feature space. An object is classified by a majority of its neighbors. K is always a positive integer. The neighbors are selected from a set of objectsfor which the correct classification is known.

The k-NN classifier is a non parametric lazy learning algorithm. A data samplein k-NN is classified on the basis of a selected number of k nearest neighbors. The assumptions followed in k-NN are

- k-NN assumes that the data is in a feature space, so they have the conceptof distance. Euclidean distance can be used to compute distance betweenvectors.
- Each training vector is associated with set of vectors and class label.
- k decides how many neighbors influence the classification.

The following rule is the majority rule that is used extensively in k-NN. The classification of the nearest neighbors can be decided by calculating the count of individual class values from all k-nearest neighbors. The class value with the majority count is classified to the sample. k is an odd number to avoid duplicate counts.

The training examples are vectors in a multidimensional feature space, each witha class label. The training phase of the algorithm consists only of storing the feature vectors and class labels of the training samples.

In the classification phase, *k* is a user-defined constant, and an unlabeled vector(a query or test point) is classified by assigning the label which is the most frequent amongthe *k* training samples nearest to that query point. Often, the classification accuracy of "k"-NN can be improved significantly if the distance metric is learned with specialized algorithms such as Large Margin Nearest Neighbour or Neighbourhood componentsanalysis.

3.5 Classification by NB

A Naive Bayes (NB) classifier is a simple probabilistic classifier based on Bayestheorem where every feature is assumed to be class-conditionally independent. A more descriptive term for the underlying probability model would be "independent feature model".

In naïve bayes learning, each instance is described by a set of features and takesa class value from a predefined set of values. Classification of instances gets difficult whenthe dataset contains a large number of features and classes because it takes enormous numbers of observations to estimate the probabilities. When a feature is assumed to be class-conditionally independent, it really means that the effect of a variable value on a given class is independent of the values of other variables.

A naive-bayes classification algorithm is a linear generative classification algorithm that applies Bayes rule and uses the independence assumption: Naive-Bayes Algorithm, NB, Naive-Bayes Learning Algorithm, Naive-Bayes, Naive-Bayes Classifier, Bayesian Classification Algorithm, Naive Bayes.3.5 Classification by NN

Multilayer perceptrons using a backpropagation algorithm are the standardalgorithm for any supervised learning pattern recognition process and the subject of ongoing research in computational neuroscience and parallel distributed processing. Theyare useful in research in terms of their ability to solve problems stochastically, which often
allows one to get approximate solutions for extremely complex

problems like fitnessapproximation.

3.6 Classification by SVM

Support Vector Machine (SVM) models have similar functional form' to neuralnetworks and radial basis functions, both of which are popular data mining techniques. SVM performs well on datasets that have many attributes, even if there are very few caseson which to train the model.

SVMs are a kind of blend of linear modelling and instance-based learning. A SVM selects a small number of critical boundary samples from each class and builds a linear discriminant function that separates them as widely as possible. In the case that no linear separation is possible, the technique of "kernel" will be used to automatically injectthe training samples into a higher-dimensional space, and to learn a separator in that space.

A support vector machine (SVM) is a hyperplane that separates two different sets of samples with maximum distance of hyperplane to nearest samples from both sets.

The formula for the output of a linear SVM is

$$u = w.\bar{x} - \bar{b}$$

In this equation w is the normal vector to the hyperplane and x is the input vector. The nearest points lie on the planes u = ±1. The distance d is

$$d = \frac{1}{\|w\|}_2$$

The maximum distance d can be expressed using optimization problem

$$\min_{w}, b \frac{1}{2} \| \overline{w} \|^2 \text{ subject to } yi(w.\overline{x_i} - \overline{b}) \geq 1,$$

where xi is the ith training sample and yi is the correct output of the SVM for the ith trainingsample. The value yi is +1 for the positive samples and -1 for the negative samples.

However, neither of these algorithms has the well-founded theoretical approachto regularization that forms the basis of SVM. The quality of generalization and ease of training of SVM is far beyond the capacities of these more traditional methods. SVM can model complex, real-world problems such as text and image classification, hand-writing recognition, and bioinformatics and bio sequence analysis. SVM performs well on datasetthat have many attributes, even if there are very few cases on which to train the model.

CHAPTER 4

DATASET DESCRIPTION

For this work, we have used the data from the database of a big airline and aircraft aviation. The application is done on five hundred data sets to compare the results.The data is given in a report layout with following sections shown in Table I. As is apparentfrom Table I, component reports have 181 attributes. The aim of the analysis is to find theattributes that affect the warning levels and a new formulation about it.

Table 4.1 Description of sample datasets used in application

Parameter name	Description
Abind	The challenging for the pilot to diagnose in flight
Aflalo	The side of the runway after landing long
Airatt	The aircraft will be put into a round out attitude shortly before it would otherwise contact the ground
Airbrot	The Burnt-out aircraft is clustered on the aft section of the flight deck,clear of the fire area
Arspd	The highest airspeed attained by an aircraft of a particular class
Aliruy	The Aligning with the runway on takeoff should be a no brainer
Altplm	The standard nominal altitude of an aircraft, in hundreds of feet
Apnopre	The pilot-interpreted make use of ground beacons and aircraft equipment such as VOR, NDB

Arhorfa	The view of heading indicator and artificial horizon after an in-flight vacuum failure
ATCerr	The service provided by ground-based controllers who direct aircrafton the ground
Autlad	The designed to make landing possible in visibility too poor to permit any form of visual landing
Autpidiso	The autopilot can control the aircraft while the pilot attends to other duties
Autpieng	The autopilot must be turned on using the Autopilot Engage Switchon the far left
Autthrot	The pilot to control the power setting of an aircraft's engines by specifying a desired flight characteristic
Enfa	Engine failure is probably your worst fear as a pilot
Lftenbrof	The left engine broke off
Rghtenbrof	The right engine broke off
Sersvib	Vibration is bad for any piece of machinery
Wnd	Effect of wind shear on aircraft trajectory

CHAPTER 5

EXPERIMENT RESULTS AND DISCUSSION

It is observed from the comparison of DT, KNN, NB, NN and SVM in the performance metric of TPRate, FPRate, Precision, Recall, F-Measure and accuracyvalues,that the results are reliable (see Table II). The J48 classifier produced the analysis of the training dataset and the classification rules. In the experiment, the phase of experiment is the evaluation and interpretation of the classification rules using the unseen data. In the experiment we have used 500 instances of the database as a training datasets. The analyseswere performed using WEKA environment. This study also proved that the NB classifierswill performance better than other classifiers on airline data. Please use only datasets which classification rules best for both on Accuracy and on classifier, as shown in Fig. 1. The performance of different classification models with ROC chart shown in figure 2.

Table 5.1 Classifiers Performance

Classifier / Performancemetric	DT	KNN	NB	NN	SVM
TP Rate	0.939	0.881	0.91	0.917	0.954
FP Rate	0.602	0.432	0.386	0.602	0.67
Precision	0.88	0.905	0.917	0.877	0.869
Recall	0.939	0.881	0.91	0.917	0.954
F-Measure	0.908	0.893	0.914	0.897	0.91
Accuracy	84.4%	82.6%	85.8 %	82.6 %	84.4%

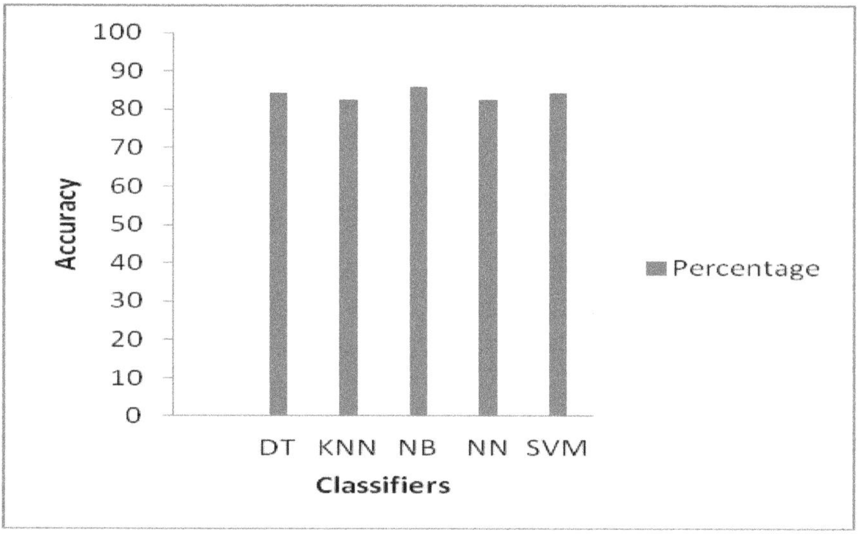

Figure 5.1 Performance of different classifiers

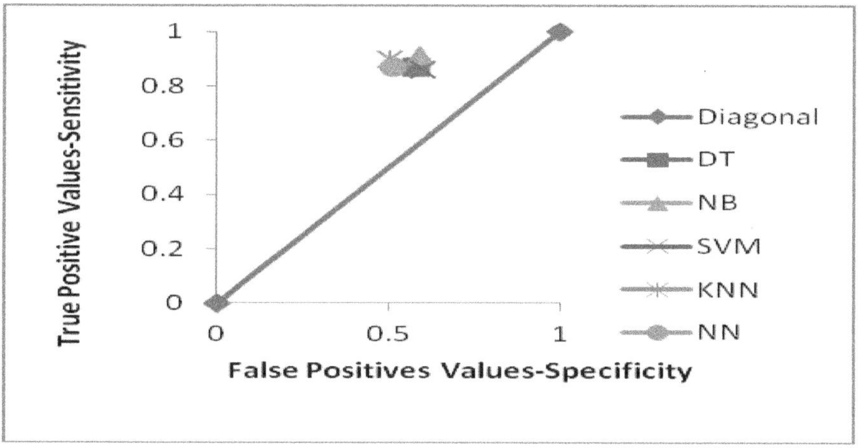

Figure 5.2 Performance of different classification Models with ROC chart

In this study we have explored the use of several classification techniques on aviation accidents data. We have discussed the basic concepts and principles of classification algorithms in data mining. The main contribution of this study is to evaluate the performance of different classification techniques such as DT, KNN, NB, NN and SVM within aviation components data.

We have applied the several feature selection techniques such as CFS, CS, GR, IG, OneR, PCA, RFF, and SU on the aircraft accident dataset. After the feature selection, we have used several classifiers such as DT, NB, SVM, KNN and NN to build classification models on the datasets with various selected subset of features. The classification models are evaluated in terms of the accuracy and other performance metric.

The classifier performance on original features is shown in Table 5.2. The Table 5.3 summarizes the classification performance in terms of accuracy for different feature selection methods. The table also displays model performance on the whole dataset. It is observed from the comparison of DT, NB, SVM, KNN and NN in the performance metric of accuracy values, that the results are reliable (see Table 5.3). The Table 5.4 displays and summarizes the comparison of classification performance in terms of average accuracy of different classifiers. The Table 5.5 shows and summarizes the average accuracy of using different feature selection methods.

All these results are mapped into a group of classifiers and features as shown in Figures 5.3 through 5.5. The multiple comparison results are presented in Figure 6.4. The figure shows the following facts: DT performed better than other classifiers and PCA methods performed better than other feature selection methods. CFS, CS, GR, OneR, IG, RFF, and SU these seven methods are ordered by their performances from worst to best. The average accuracy of classifiers results is presented in Figure 5.4. The Average accuracy of different classifiers on reduced features by different feature selection methods results is presented in Figure 5.5.

The ROC (receiver operating characteristic curve) is a graphical plot which illustrates the performance of a binary classifier as its discrimination threshold is varied. It is created by plotting the fraction of true positives of out of the positives (TPR = true positive rate) vs. the fraction of false positives out of the negatives (FPR = false positive rate), at various threshold settings. TPR is also known as sensitivity (also called recall in some fields), and FPR is one minus the specificity or true negative rate. Figure 5.6 classifier performances on original features with ROC curve. Figure 5.7 shows classifier performances on reduced features with ROC curve

This study also proved that the DT classifier would show better performance than other classifiers and PCA method would perform better than other feature selection methods on airline data.

Table 5.2 Classifiers performance on original features

Classifiers	Correctly Classified Instances	Incorrectly Classified Instances	Accuracy (%)
DT	422	78	84.4
NB	429	71	85.8
SVM	422	78	84.4
KNN	413	87	82.6
NN	413	87	82.6

Table 5.3 Classifiers performance of reduced features

Feature Selection Method	DT	NB	SVM	KNN	NN	AVG
CFS	97.08	97.08	97.08	97.08	97.08	97.08%
CS	91.08	90.02	91.06	91.02	86.06	90.28%
GR	99.04	99.04	99.04	99.04	99.04	99.04%
IG	99.04	99.04	99.04	99.04	98.04	99.02%
OneR	98.04	98.04	98.04	98.00	97.06	98.16%
PCA	99.08	99.08	99.08	99.06	99.08	99.76%
RF	95.04	93.08	95.04	94.08	91.04	94.16%
SU	99.04	99.04	99.04	94.04	93.00	97.12%
AVG	97.68%	97.28%	97.65%	96.83%	95.05%	

Table 5.4 Average accuracy of different classifiers on reduced features

Classifiers	Accuracy (%)
Decision Tree	**97.68**
Naive Bayes	97.28
Support Vector Machines	97.65
K-Nearest Neighbor	96.83
Neural Network	95.05

Table 5.5 Average accuracy of different feature selection method on different classifiers

Feature Selection Methods	Accuracy (%)
CFS Subset evaluation	97.08
Consistency subset evaluation	90.28
Gain ratio attribute evaluation	99.04
Info gain attribute evaluation	99.02
OneR attribute evaluation	98.16
Principal Component Analysis	**99.76**
Relief attribute evaluation	94.16
Symmetric Uncertainty	97.12

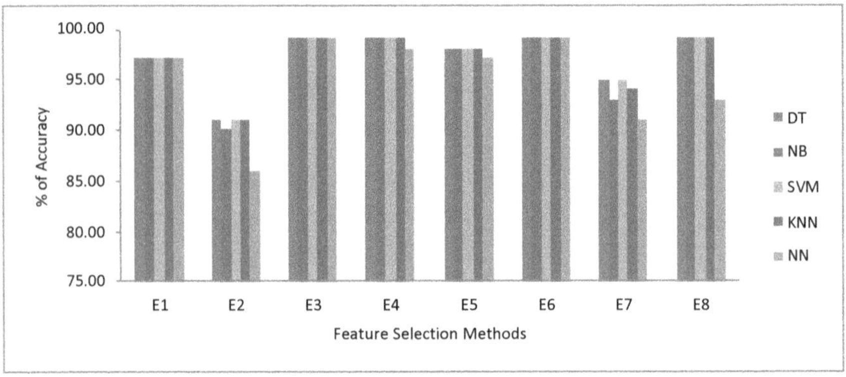

Figure 5.3 Classifier performances on reduced features

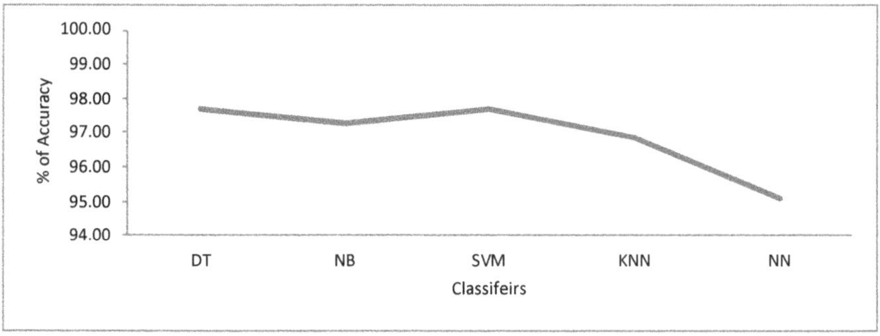

Figure 5.4 Average accuracy of different classifiers on original features

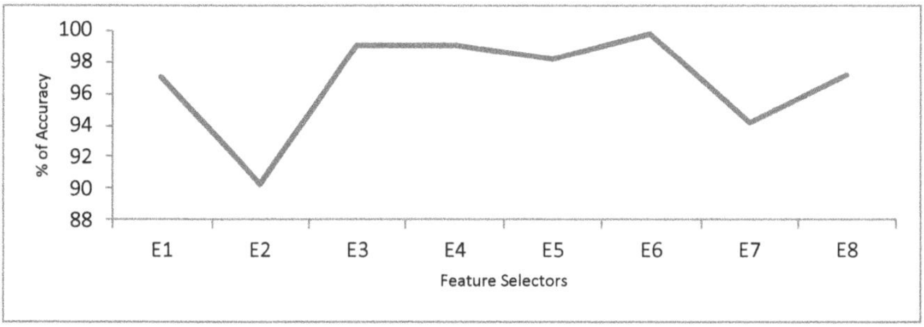

Figure 5.5 Average accuracy of different classifiers on reduced features by different feature selection methods

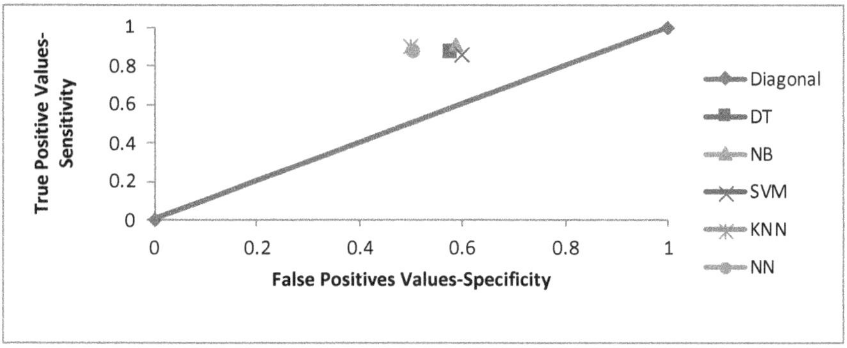

Figure 5.6 Classifier performances on original features with ROC curve

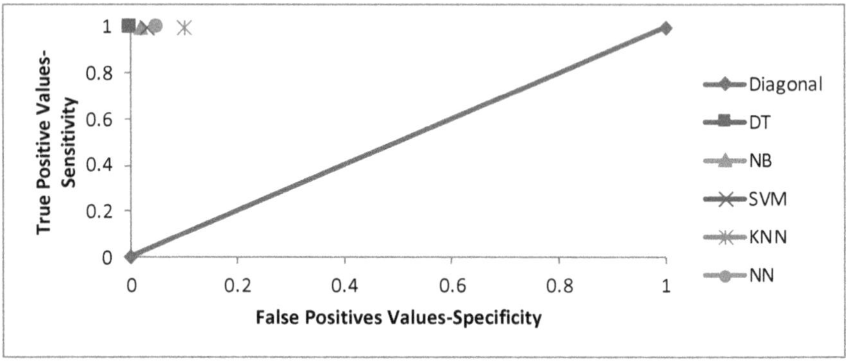

Figure 5.7 Classifier performances on reduced features with ROC curve

The accuracy of decision tree classifier was improved from 84.4 % to 97.68% where 84.4% is the accuracy with original features and 97.68% is the average accuracy of applying different feature selectors.

The accuracy of Naive bayes classifier was improved from 85.8 % to 97.28% where 85.8 % is the accuracy with original features and 97.28% is the average accuracy of applying different feature selectors.

The accuracy of support vector machine classifier was improved from 84.4 % to 97.65% where 84.4% is the accuracy with original features and 97.65% is the average accuracy of applying different feature selectors.

The accuracy of K-nearest neighbour classifier was improved from 82.6 % to 96.83% where 82.6% is the accuracy with original features and 96.83% is the average accuracy of applying different feature selectors.

The accuracy of artificial neural network classifier was improved from 82.6 % to 95.05% where 82.6% is the accuracy with original features and 95.05% is the average accuracy of applying different feature selectors.

The decision tree shows significant detection performance on selected feature by different feature selectors than other classifiers and the performance of principal component analysis for feature reduction is comparatively better than other methods.

CHAPTER 6

CONCLUSION

In this work we have explored the use of different classification techniques on aviation accidents data. We have discussed the basic concepts and principles of classification algorithms in data mining. The main contribution of this study is to evaluatethe performance of different classification techniques are DT, KNN, NB, NN and SVM within aviation components data. We talked about the NB rules that were found by this analysis with the experts of Aviation Company better than other classifiers. As a result saidthat NB rules are some safety recommendations are address to the Airline Aviation Administration.

REFERENCES

[1] A.S. Chang, S.S. Leu, Data mining model for identifying project profitability variables,International Journal of Project Management 24 (2006) 199–206.
[2] C. Apte, S. Weiss, Data mining with decision trees and decision rules, Future GenerationComputer Systems (1997).
[3] C.C. Chang, R.S. Chen, Using data mining technology to solve classification problems,A Case Study of Campus Digital Library, Institute of Information Management, National Chiao Tung University, Hsinchu, Taiwan, 2006.
[4] C.K. Mai, M. Krishna, A.V. Reddy, PolyAnalyst Application for Forest Data Mining, IEEE, 2005.
[5] Crone, S. F., Lessmann, S., & Stahlbock, R. (2006). The impact of pre-processing on data mining: An evaluation of classifier sensitivity in direct marketing. European Journal of Operational Research, 173, 781–800.
[6] E. Kirkos, C. Spathis, Y. Manolopoulos, Data mining techniques for the detection of fraudulent financial statements, Expert Systems with Applications 32 (2007) 995–1003.
[7] Editorial of Engineering Applications of Artificial Intelligence 19, Recent Advances inData Mining, 2006, pp. 361–362.
[8] F. Emekci, O.D. Sahin, D. Agrawal, El Abbadi, Privacy preserving decision tree learning over multiple parties, Data and Knowledge Engineering 63 (2007) 348–361.
[9] Feyza Gürbüz, Lale Özbakir, Hüseyin Yapici (2009). Classification rule discovery for the aviation incidents resulted in fatality. Knowledge-Based Systems 22 (2009) 622– 632.
[10] Feyza Gürbüz, Lale Özbakir , Hüseyin Yapici (2011). Data mining and pre-processing application on component reports of an airline company in Turkey. Expert Systems with Applications 38 (2011) 6618–6626.
[11] G.K.F. Tso, K.K.W. Yau, Predicting Electricity Energy Consumption: A Comparison of Regression Analysis, Decision Tree and Neural Networks, 2005,
<www.sciencedirect.com>.
[12] H.J. Shyur, A quantitative model for aviation safety risk assessment, Computers andIndustrial Engineering (2007).

[13] Hand, D., Manila, H., & Smyth, P. (2001). Principles of data mining. Londan, England:Cambridge Massachusetts.
[14] Hu, X. (2003). DB-reduction: A data pre-processing algorithm for data mining applications. Applied Mathematics Letters, 16, 889–895.
[15] Introduction to Data Mining and Knowledge Discovery, Two Crows Corporation, 3rd ed., 1999.
[16] M. Bineid, J.P. Fielding, Development of a civil aircraft dispatch reliability prediction methodology, Aircraft Engineering and Aerospace Technology 75 (6) (2003) 588–594.
[17] M.J. Aitkenhead, A co-evolving decision tree classification method, Expert Systems with Applications 34 (2006) 18–25.
[18] Nazeri, Z., & Jianping, Z. (2002). Mining aviation data to understand impacts of severeweather on airspace system performance. In Proceedings of the international conferenceon information technology. IEEE.
[19] S. Dessureault, A. Sinuhaji, P. Coleman, Data mining mine safety data, Mining Engineering Littleton 59 (8) (2007) 64. 7 pgs.
[20] T.C. Hsia, A.J. Shie, L.C. Chen, Course Planning of Extension Education to Meet Market Demand by Using Data Mining Techniques-An Example of Chinkuo Technology University in Taiwan, Taiwan, 2006.

www.ingramcontent.com/pod-product-compliance
Lightning Source LLC
LaVergne TN
LVHW050923210926
848482LV00012B/516
* 9 7 8 1 9 2 2 6 1 7 6 5 1 *